❧ This Book Belongs To ❧

Name :

Phone :

Address :

Project Name :

Project No :

Date :

Foreman :

Day :

Visitors

Schedule

Problems

Safety Issues

Summary Of Work

Signature :

Employee	Trade	Hours	Overtime

Equipment On Site	No. Units

Materials Delivered	No. Units	Equipment Rented	Rate

Others

Notes :

Project Name :

Foreman :

Project No :

Date :

Day :

Visitors	Schedule

Problems	Safety Issues

Summary Of Work

Signature :

Employee	Trade	Hours	Overtime

Equipment On Site	No. Units

Materials Delivered	No. Units	Equipment Rented	Rate

Others

Notes :

Project Name :

Project No :

Date :

Foreman :

Day :

Visitors

Schedule

Problems

Safety Issues

Summary Of Work

Signature :

Employee	Trade	Hours	Overtime

Equipment On Site	No. Units

Materials Delivered	No. Units	Equipment Rented	Rate

Others

otes :

Project Name :

Foreman :

Project No :

Date :

Day :

Visitors	Schedule

Problems	Safety Issues

Summary Of Work

Signature :

Employee	Trade	Hours	Overtime

Equipment On Site	No. Units

Materials Delivered	No. Units	Equipment Rented	Rate

Others

Notes :

Project Name :

Foreman :

Project No :

Date :

Day :

Visitors	Schedule

Problems	Safety Issues

Summary Of Work

Signature :

Employee	Trade	Hours	Overtime

Equipment On Site	No. Units

Materials Delivered	No. Units	Equipment Rented	Rate

Others

Notes :

Project Name :

Project No :

Date :

Foreman :

Day :

Visitors

Schedule

Problems

Safety Issues

Summary Of Work

Signature :

Employee	Trade	Hours	Overtime

Equipment On Site	No. Units

Materials Delivered	No. Units	Equipment Rented	Rate

Others

otes :

Project Name : | Project No :

Date :

Foreman : | Day :

Visitors

Schedule

Problems

Safety Issues

Summary Of Work

Signature :

Employee	Trade	Hours	Overtime

Equipment On Site	No. Units

Materials Delivered	No. Units	Equipment Rented	Rate

Others

Notes :

Project Name :

Project No :

Date :

Foreman :

Day :

Visitors

Schedule

Problems

Safety Issues

Summary Of Work

Signature :

Employee	Trade	Hours	Overtime

Equipment On Site	No. Units

Materials Delivered	No. Units	Equipment Rented	Rate

Others

Notes :

Project Name :

Foreman :

Project No :

Date :

Day :

Visitors

Schedule

Problems

Safety Issues

Summary Of Work

Signature :

Employee	Trade	Hours	Overtime

Equipment On Site	No. Units

Materials Delivered	No. Units	Equipment Rented	Rate

otes :

Project Name :

Foreman :

Project No :

Date :

Day :

Visitors

Schedule

Problems

Safety Issues

Summary Of Work

Signature :

Employee	Trade	Hours	Overtime

Equipment On Site	No. Units

Materials Delivered	No. Units	Equipment Rented	Rate

Others

Notes :

Project Name :

Foreman :

Project No :

Date :

Day :

Visitors

Schedule

Problems

Safety Issues

Summary Of Work

Signature :

Employee	Trade	Hours	Overtime

Equipment On Site	No. Units

Materials Delivered	No. Units	Equipment Rented	Rate

Others

Notes :

Project Name :

Project No :

Date :

Foreman :

Day :

Visitors	Schedule

Problems	Safety Issues

Summary Of Work

Signature :

Employee	Trade	Hours	Overtime

Equipment On Site	No. Units

Materials Delivered	No. Units	Equipment Rented	Rate

Others

Notes :

Project Name :

Project No :

Date :

Foreman :

Day :

Visitors

Schedule

Problems

Safety Issues

Summary Of Work

Signature :

Employee	Trade	Hours	Overtime

Equipment On Site	No. Units

Materials Delivered	No. Units	Equipment Rented	Rate

Others

Notes :

Project Name :

Foreman :

Project No :

Date :

Day :

Visitors

Schedule

Problems

Safety Issues

Summary Of Work

Signature :

Employee	Trade	Hours	Overtime

Equipment On Site	No. Units

Materials Delivered	No. Units	Equipment Rented	Rate

<table>
<tr><td align="center">Others</td></tr>
</table>

Notes :

Project Name :

Foreman :

Project No :

Date :

Day :

Visitors

Schedule

Problems

Safety Issues

Summary Of Work

Signature :

Employee	Trade	Hours	Overtime

Equipment On Site	No. Units

Materials Delivered	No. Units	Equipment Rented	Rate

otes :

Project Name : | Project No :

Date :

Foreman : | Day :

Visitors

Schedule

Problems

Safety Issues

Summary Of Work

Signature :

Employee	Trade	Hours	Overtime

Equipment On Site	No. Units

Materials Delivered	No. Units	Equipment Rented	Rate

Others

Notes :

Project Name :

Project No :

Date :

Foreman :

Day :

Visitors	Schedule

Problems	Safety Issues

Summary Of Work

Signature :

Employee	Trade	Hours	Overtime

Equipment On Site	No. Units

Materials Delivered	No. Units	Equipment Rented	Rate

Others

Notes :

Project Name :

Project No :

Date :

Foreman :

Day :

Visitors	Schedule

Problems	Safety Issues

Summary Of Work

Signature :

Employee	Trade	Hours	Overtime

Equipment On Site	No. Units

Materials Delivered	No. Units	Equipment Rented	Rate

Others

otes :

Project Name :

Project No :

Date :

Foreman :

Day :

Visitors

Schedule

Problems

Safety Issues

Summary Of Work

Signature :

Employee	Trade	Hours	Overtime

Equipment On Site	No. Units

Materials Delivered	No. Units	Equipment Rented	Rate

Others

Notes :

Project Name :

Project No :

Date :

Foreman :

Day :

Visitors

Schedule

Problems

Safety Issues

Summary Of Work

Signature :

Employee	Trade	Hours	Overtime

Equipment On Site	No. Units

Materials Delivered	No. Units	Equipment Rented	Rate

Others

Notes :

Project Name :

Project No :

Date :

Foreman :

Day :

Visitors	Schedule

Problems	Safety Issues

Summary Of Work

Signature :

Employee	Trade	Hours	Overtime

Equipment On Site	No. Units

Materials Delivered	No. Units	Equipment Rented	Rate

Others

otes :

Project Name :

Foreman :

Project No :

Date :

Day :

Visitors	Schedule

Problems	Safety Issues

Summary Of Work

Signature :

Employee	Trade	Hours	Overtime

Equipment On Site	No. Units

Materials Delivered	No. Units	Equipment Rented	Rate

Others

Notes :

Project Name :

Project No :

Date :

Foreman :

Day :

Visitors	Schedule

Problems	Safety Issues

Summary Of Work

Signature :

Employee	Trade	Hours	Overtime

Equipment On Site	No. Units

Materials Delivered	No. Units	Equipment Rented	Rate

Others

Notes :

Project Name :

Project No :

Date :

Foreman :

Day :

Visitors

Schedule

Problems

Safety Issues

Summary Of Work

Signature :

Employee	Trade	Hours	Overtime

Equipment On Site	No. Units

Materials Delivered	No. Units	Equipment Rented	Rate

otes :

Project Name :

Project No :

Date :

Foreman :

Day :

Visitors

Schedule

Problems

Safety Issues

Summary Of Work

Signature :

Employee	Trade	Hours	Overtime

Equipment On Site	No. Units

Materials Delivered	No. Units	Equipment Rented	Rate

Others

Notes :

Project Name :

Project No :

Date :

Foreman :

Day :

Visitors

Schedule

Problems

Safety Issues

Summary Of Work

Signature :

Employee	Trade	Hours	Overtime

Equipment On Site	No. Units

Materials Delivered	No. Units	Equipment Rented	Rate

Others

Notes :

Project Name :   Project No :

Date :

Foreman :   Day :

Visitors

Schedule

Problems

Safety Issues

Summary Of Work

Signature :

Employee	Trade	Hours	Overtime

Equipment On Site	No. Units

Materials Delivered	No. Units	Equipment Rented	Rate

Others

otes :

Project Name :

Foreman :

Project No :

Date :

Day :

Visitors	Schedule

Problems	Safety Issues

Summary Of Work

Signature :

Employee	Trade	Hours	Overtime

Equipment On Site	No. Units

Materials Delivered	No. Units	Equipment Rented	Rate

<table>
<tr><th>Others</th></tr>
</table>

Notes :

Project Name :

Project No :

Date :

Foreman :

Day :

Visitors

Schedule

Problems

Safety Issues

Summary Of Work

Signature :

Employee	Trade	Hours	Overtime

Equipment On Site	No. Units

Materials Delivered	No. Units	Equipment Rented	Rate

Others

Notes :

Project Name :

Project No :

Date :

Foreman :

Day :

Visitors

Schedule

Problems

Safety Issues

Summary Of Work

Signature :

Employee	Trade	Hours	Overtime

Equipment On Site	No. Units

Materials Delivered	No. Units	Equipment Rented	Rate